KEW POCKETBOOKS

MEXICAN PLANTS

Curated by Brie Langley and Lydia White

Kew Publishing
Royal Botanic Gardens, Kew

THE LIBRARY AND ARCHIVES AT KEW

KEW HOLDS ONE OF THE LARGEST COLLECTIONS of botanical literature, art and archive material in the world. The library comprises 185,000 monographs and rare books, around 150,000 pamphlets, 5,000 serial titles and 25,000 maps. The Archives contain vast collections relating to Kew's long history as a global centre of plant information and a nationally important botanic garden including 7 million letters, lists, field notebooks, diaries and manuscript pages.

The Illustrations Collection comprises 200,000 watercolours, oils, prints and drawings, assembled over the last 200 years, forming an exceptional visual record of plants and fungi. Works include those of the great masters of botanical illustration such as Ehret, Redouté, the Bauer brothers, Thomas Duncanson, George Bond and Walter Hood Fitch. Our special collections include historic and contemporary originals prepared for *Curtis's Botanical Magazine*, the work of Margaret Meen, Thomas Baines, Margaret Mee, Joseph Hooker's Indian sketches, Edouard Morren's bromeliad paintings, 'Company School' works commissioned from Indian artists by Roxburgh, Wallich, Royle and others, and the Marianne North Collection, housed in the gallery named after her in Kew Gardens.

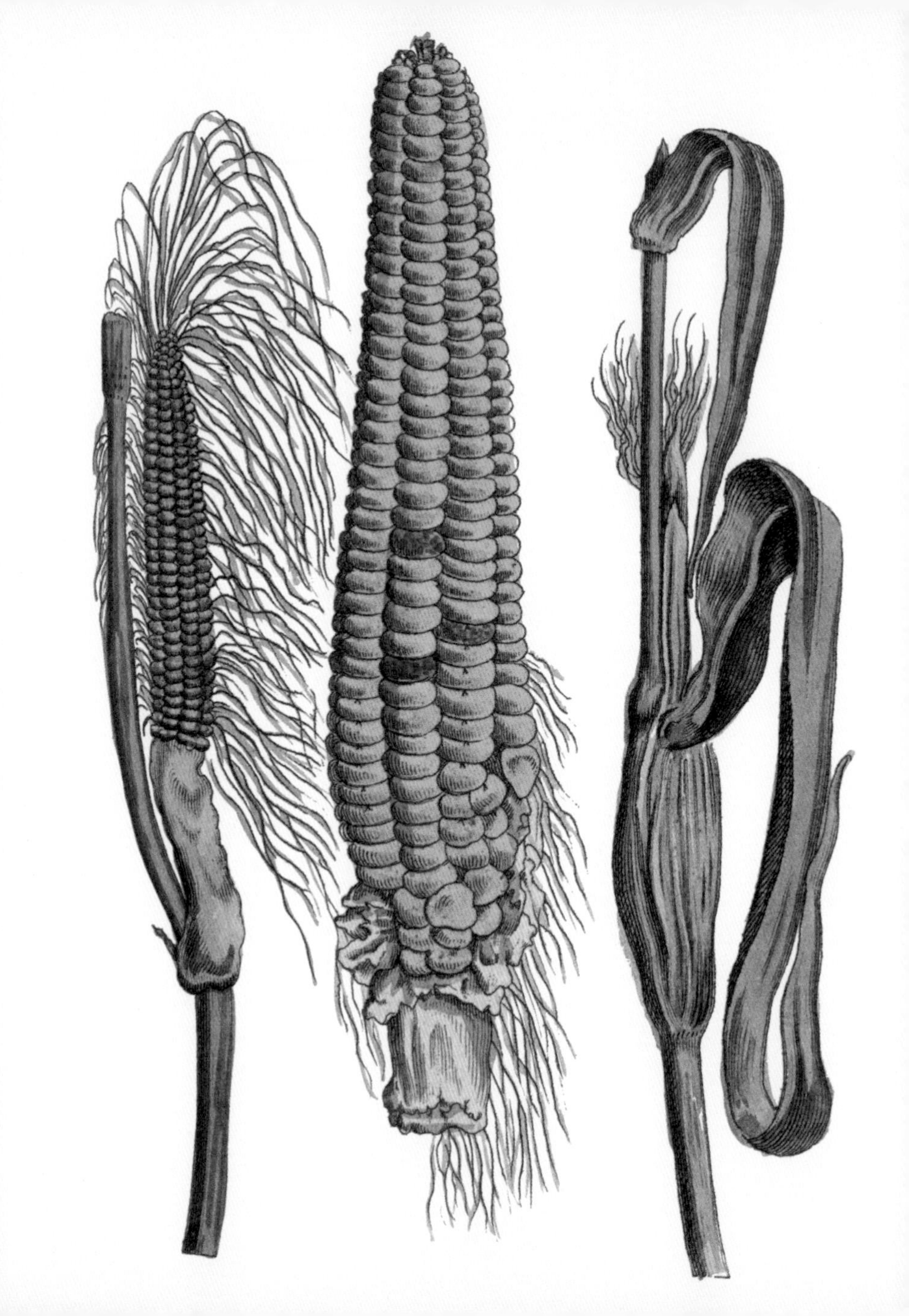

INTRODUCTION

MEXICO IS A COUNTRY OF VIBRANT FOOD, music, history and passion. But plants? The traditional image of a lone saguaro cactus in the dusty desert comes to mind, years of media and film burned into our subconscious. This is of course far from the truth.

Mexico is a biodiversity hotspot, representing 12% of global biodiversity. There are many different reasons why this may be the case, including the country's position as a physical connection between North and South America. Fauna and flora would have migrated through the country, and still do, with some deciding to stay rather than move on. The sheer number of environments, habitats and niches that are available in this amazing country is astounding. And this is what biodiversity needs. It loves a good niche.

Even very generalist plants will succumb to the evolutionary pull that is natural selection. Gradually, gradually, slowly but surely, they become more specialised, more adept at dealing with that one situation while disregarding all others. Soon, they can no longer mate with those they used to call kin, and a new species is born. Other plants follow suit, and so the biodiversity of the area increases. With more species comes more competition, which leads to yet more specialisation in order to escape that pressure. But this is not a continuous vicious cycle. Plants will often work together within their highly specialised communities; they may share nutrients, warn others of attack or even provide the very habitat that another plant needs to survive. Biodiversity is a strength and will allow plant communities to cope better with challenges, such as climate change, by working together as a network.

Mexican geography is one of great contrast, providing many different environments for plant communities to specialise in. These diverse ecosystems range from the dry heat of the Sonoran and Chihuahuan deserts of the North, peppered with saguaro and prickly pear cactus, to the permanent snow cover of Citlaltéptl, Mexico's highest peak and one of its many volcanos. Cool broadleaf and conifer forests form the backbone of the highlands, where monarch butterflies come to rest; while the heavy-laden tropical rainforests of the south, dripping in orchids and bromeliads, teem with rare plants only found here. And the miles and miles of coastline, 5,780 miles (9,300 km) in fact, of sand, rock and cliff that rise from the Gulf of Mexico to the east and the Pacific Ocean to the west. It is easy to see how 26,000 different species of flowers could call Mexico home.

At Kew, we grow many of the more tender Mexican plants in our glasshouses. The Temperate House, bright and airy with a Mediterranean climate, has a whole Mexican section. Towering stands of *Dahlia imperialis* flank the central path of the house, their graceful hollow stems bow softly in the breeze. The stems were once used by the Aztec to transport water as they are lightweight, but incredibly tough once cured. The eldest and smallest of our glasshouses is the Palm House, dedicated solely to the rainforest ecosystem. Here we grow much of our cycad collection, many of which are endemic, or native only to, Mexico. This ancient group of plants are now a cause of concern to conservationists as overcollection threatens many with extinction – so always be sure of your sources, especially if buying plants online. In the Palm House we run a cycad phenology programme, a bit like a breeding programme in a zoo, to produce seed for the next generation and to share with other gardens around the world.

Kew is currently working with partners in Mexico – scientists, horticulturalists and everyone in between – to help conserve Mexican flora *in situ*. Already, the effects of climate change are being felt on many of Mexico's diverse ecosystems. In the cloud forests of Veracruz, climate change and deforestation has led to a significant loss of tree cover. Four hundred different species of tree identified as important to local communities will be grown by local growers to reforest areas lost to farming and logging. The change in climate is also affecting traditional farming methods. A *milpa*, the anglicised 'Three Sisters,' is a combination of maize, beans and squash grown together in one space, each supporting and benefitting each other. In certain areas, lack of rain and a shorter growing season has greatly reduced the productivity of this intercrop arrangement. Kew is working with partners in the area to develop a way forward, either through plant breeding or sourcing new areas to farm.

Humans and plants have always been inextricably linked, sometimes totally dependent on the other. Maize, for example, only exists because it was bred from wild teosinte corn over 5,000 years ago. Over millennia, many plants have been used by the native peoples of Mexico, including the Aztec and Maya, for medicine, food or even as status symbols. In modern times, some are globally important food crops, whilst others are being investigated for cures for cancer. Still more are prized garden plants, though some of them need that extra input to get them through Britain's wet winters.

Dahlias fall into this category, I find. Throughout the book I have tried to add some growing tips, where I can, to help bring that touch of Mexican sunshine to your garden at home.

So, from chocolate to tequila, dahlia to cosmos – let this little book serve as an introduction to the great diversity and beauty of Mexican flora.

Brie Langley
Botanical Horticulturist, the Palm House
Royal Botanic Gardens, Kew

THE CAVE O

UERNAVACA

Nº 1

Tillandsia ionantha

airplant
by Walter Hood Fitch from
Curtis's Botanical Magazine, 1847

These plants are covered in so many tiny hairs that they look grey rather than green. The hairs allow *Tillandsia* to gather nutrients and moisture from the air instead of needing roots, which helps them to grow in otherwise inhospitable places. Some even grow on powerlines.

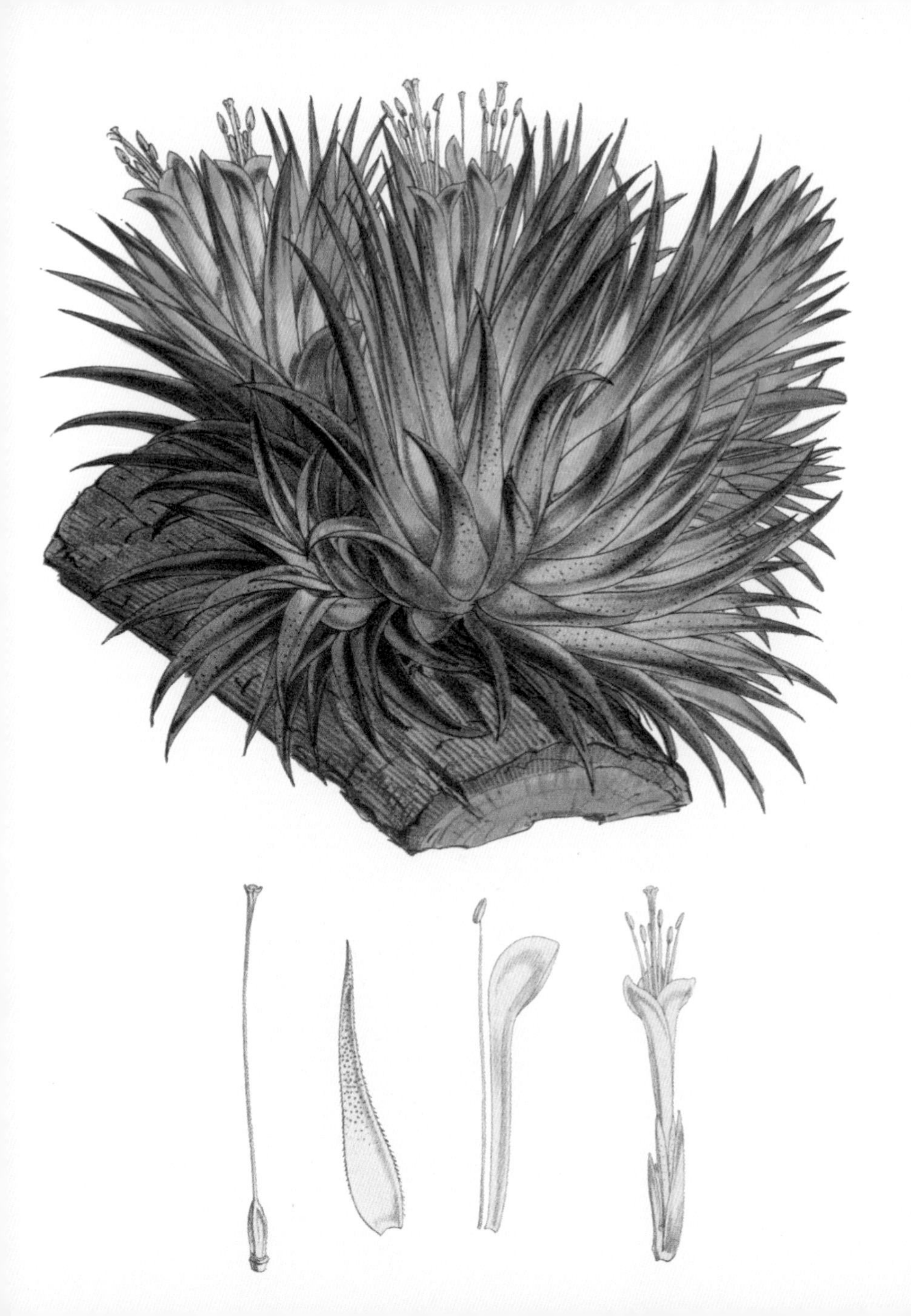

N° 2

Theobroma cacao

cacao, cocoa tree
by A. Faguet from Henri Baillon
Dictionnaire de Botanique, 1876–92

It has long been thought that this stout shrub depended upon only one species of gnat to pollinate its flowers, yet many cocoa pods have been produced at Kew with no intervention. Either the chocolate tree is not as fussy as we thought; or the tenacity of our gnats is more so.

Nº 3

Argemone mexicana

Mexican poppy
by Rungia from Robert Wight
Illustrations of Indian Botany, 1840–50

This plant is incredibly successful at colonising new ground and its efforts have now made it a weed in many parts of the world. It has been used traditionally as a purgative and is being investigated as a biofuel.

No 4

Laelia superbiens

by Sarah Anne Drake from James Bateman
The Orchidaceae of Mexico and Guatemala,
1837–43

This spectacular orchid can produce a flowering stem up to 2 m (6½ ft) tall. The bright purple flowers are arranged in a cluster at the top of the stem and are each 12 cm (4½ in) wide.

Nº 5

Abies religiosa

sacred fir, oyamel fir
from Kew Collection, 1840–80

From December until March, these giant conifers are covered in heavy swags of dark orange and brown – huge colonies of hibernating monarch butterfly. There are only a few locations where the migrating butterfly comes to rest. Some are now threatened by logging.

Nº 6

Sterculia mexicana

sterculia, tropical chestnut
by Marianne North from the
Marianne North Collection, Kew, 1880–90

A large tree with furry leaves that look a little like a chestnut tree. But don't be tempted! The fruit is covered in small irritant hairs.

Nº 7

Bixa orellana

achiote

from Friedrich Gottlob Hayne
Getreue Darstellung und Beschreibung der in der Arzneykunde Gebräuchlichen Gewächse, 1805–46

This tree is the source of annatto, which comes from the red outer covering of its seeds. Not only is this used as a natural food colouring, dye and even lipstick; the sauce made from annatto forms the basis of many traditional dishes.

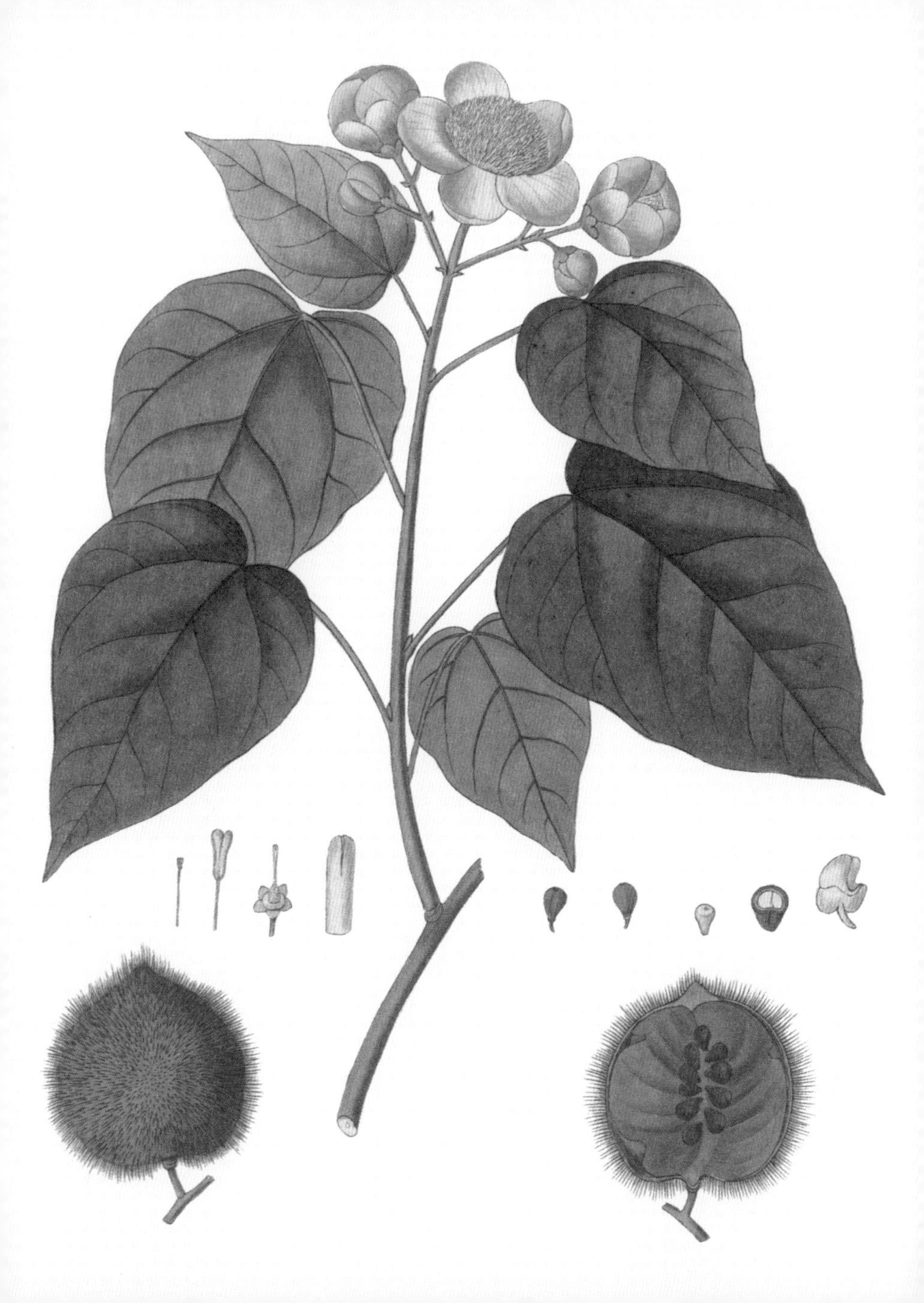

N° 8

Tagetes erecta

marigold, flor de muerto, African marigold
from Shirley Hibberd
Familiar Garden Flowers, 1879–87

Although a common sight in gardens throughout the world, this plant holds special status in its native Mexico. It has been used ceremonially for thousands of years and is now strongly associated with the Day of the Dead. The flowers are edible, provide a strong yellow pigment and may have many health benefits, including purging the body of worms.

No 9

Chamaedorea geonomiformis

chamaedorea
by Walter Hood Fitch from
Curtis's Botanical Magazine, 1874

An understory palm with almost disproportionately large leaves; this elegant, single stemmed species may eventually reach 2 m (6½ ft) in height.

Nº 10

Smilax aristolochiifolia

Mexican sarsaparilla
from F. E. Köhler
Köhler's Medizinal-Pflanzen, 1887

There are many plants used around the world to create sarsaparilla, a kind of root beer that was traditionally used as a health tonic. *Smilax aristolochiifolia* has many medicinal properties and may even treat cancer, yet this has not been fully analysed.

Nº 11

Zea mays

corn, maize

by Otto Wilhelm Thome from *Flora von Deutschland, Osterreich und der Schweiz*, 1886–89

Maize was first cultivated in Tehuacán around 3600 BCE and is now a globally important food crop. It uses wind to move pollen from its upper male flowers to the female, corn-bearing, flowers below. If you want to try growing it at home, plant in a block rather than a line to increase your yield of corn on the cob.

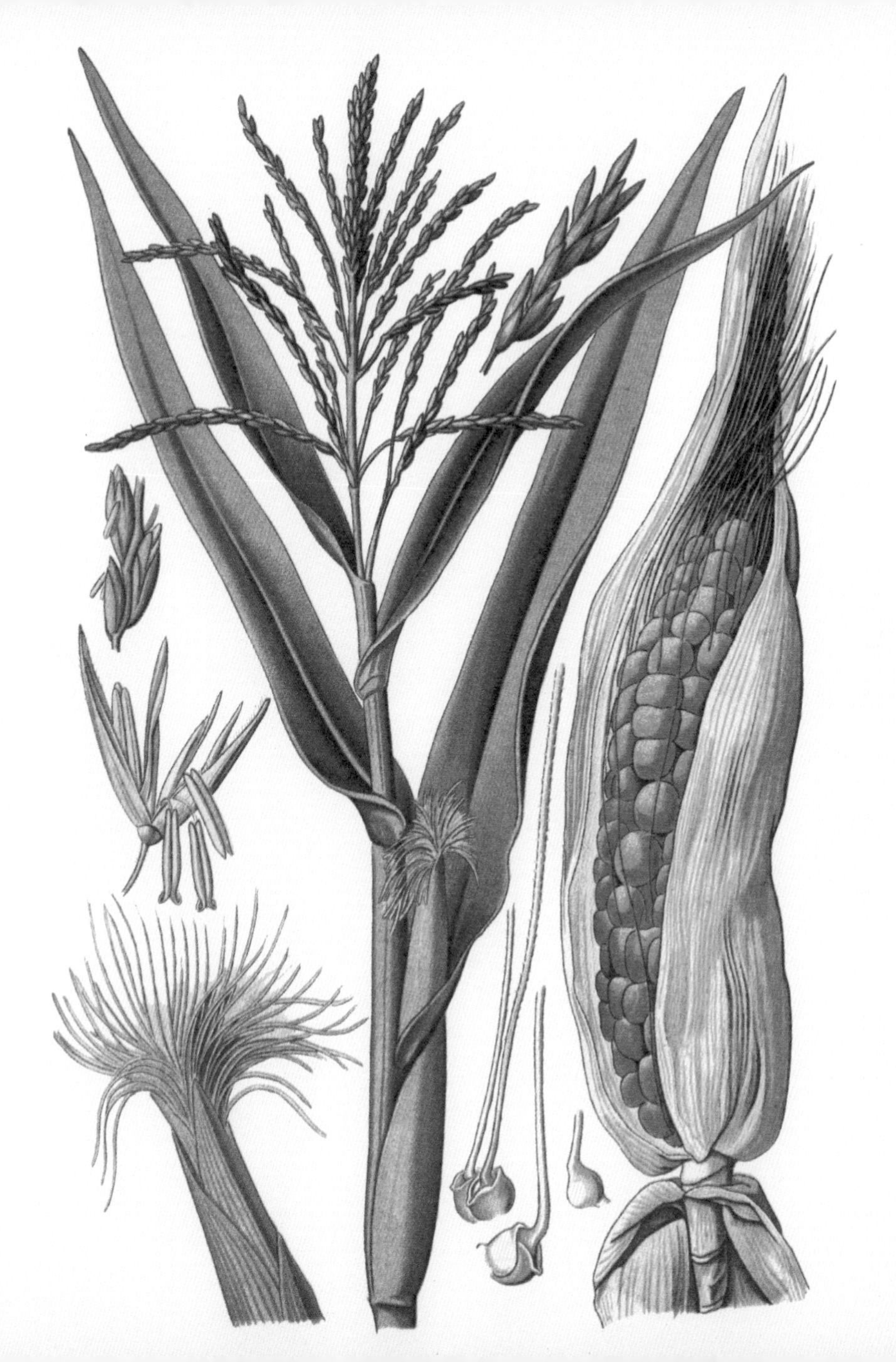

Nº 12

Tigridia pavonia

maravilla, Mexican shellflower
by Matilda Smith from
Curtis's Botanical Magazine, 1889

These brilliant blooms are named after their 'tiger-like' markings and were once eaten by the indigenous peoples of Mexico; the roasted bulb is said to have a chestnut flavour. Although they look highly specialised, these bulbs are very easy to grow at home and will come back year after year.

Nº 13

Caesalpinia pulcherrima

peacock flower, Barbados pride
from Nikolaus Joseph von Jacquin *Selectarum Stirpium Americanarum Historia*, 1780–1

An incredibly beautiful shrub. The flowers almost have the appearance of butterflies and are pollinated by hummingbirds. A large flat seed pod is then produced, just like a bean, but the seeds inside are inedible.

Nº 14

Echeveria agavoides

lipstick echeveria

from L. Van Houtte *Flore des Serres et des Jardins de l'Europe,* 1873

Now a popular house plant, this succulent can be found in rocky outcrops throughout Mexico. Succulents store water in their leaves to cope with long periods of drought or irregular rainfall. They can also propagate themselves from their leaves alone, so do not worry if a healthy leaf falls off at home. Lie it on a sunny windowsill and it will soon produce roots.

Nº 15

Ageratum houstonianum

flossflower, blueweed
by E. D. Smith from Robert Sweet
The British Flower Garden, 1823–38

These blue pom-pom flowers may be native to pasture and grassland, but they are toxic to grazing animals and have now become an agricultural weed in many countries. *Ageratum* is popularly used as a garden plant.

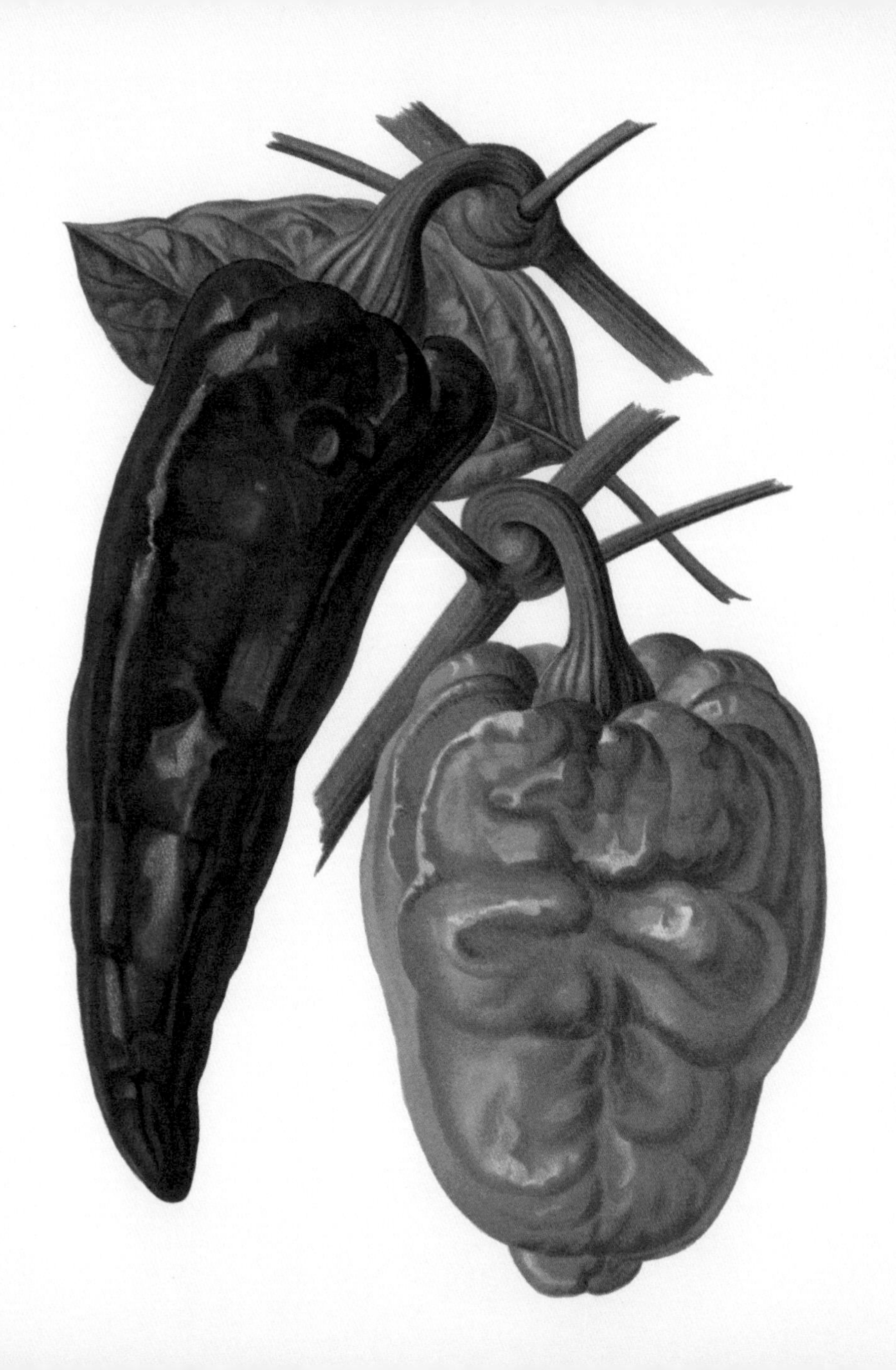

Nº 16

Capsicum annuum

chilli pepper, pimiento, chiplín
by P. de Longpré from E. A. Carrière
Revue Horticole, 1889

---

42 cultivars have been developed from this one species, including bell peppers and jalapeños. Whether spicy or sweet, capsicum serves as the basis for a huge array of traditional dishes and is now seen as a fundamental ingredient world-wide.

No 17

Cosmos bipinnatus

Mexican aster
by Sydenham Teast Edwards from
Curtis's Botanical Magazine, 1813

Much loved throughout the world as a garden plant, cosmos is native to Mexico, Costa Rica and Guatemala. The flowers are often visited by butterflies and bees, but the plant's success has made it a weed in some countries.

Nº 18

Opuntia decumbens

prickly pear
by Walter Hood Fitch from
Curtis's Botanical Magazine, 1847

The fruit and leaves of the prickly pear cactus feature in many traditional dishes and the plant is also a source of betanin, a red dye. The dye cochineal is produced by crushing scale insects that live on this plant.

Nº 19

Gymnosperma glutinosum

gumhead, zotla, popote, popotillo, tatalencho
from Sydenham Teast Edwards
Edwards' Botanical Register, 1815–47

Used traditionally to treat diarrhoea, this plant is currently being investigated for its anti-microbial and anti-fungal properties.

Nº 20

Tithonia rotundifolia

Mexican sunflower, mammoth torch flower
by T. Glover from
Curtis's Botanical Magazine, 1834

The mammoth torch flower is very easy to grow from seed and will quickly reach over 2 m (6½ ft) tall in the right situation. This drought tolerant plant can be used both as an annual hedge and green manure; whereby plants are grown to be churned into the earth again to give nutrients to the soil.

Nº 21

Sicyos edulis

chayote

from Nikolaus Joseph von Jacquin *Selectarum Stirpium Americanarum Historia,* 1780–1

The chayote is a very mild tasting gourd that can be eaten raw, but is mostly cooked to soften the texture. All parts of the plant can be eaten, the tubers could be used as a potato substitute.

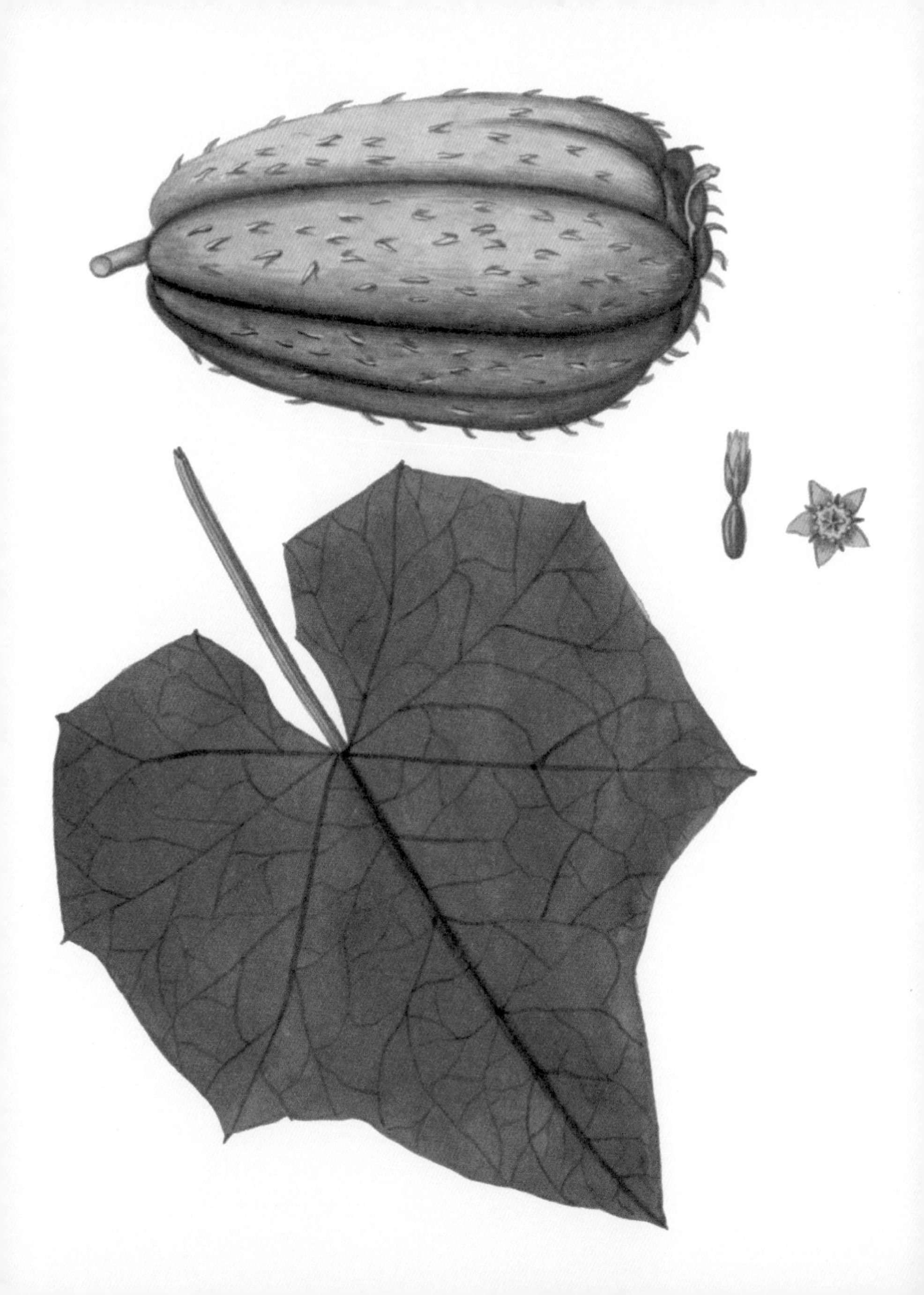

N° 22

Physalis philadelphica

tomatillo
from Joseph Franz von Jacquin
Eclogae Plantarum Rariorum aut Minus Cognitarum, 1811–44

The tomatillo is a small, green tomato-like fruit that is often eaten raw and was regarded as greater than the standard tomato by the Aztec and Maya. Tomatillo forms the basis of many dishes to this day and can be used at different stages of ripeness to suit how tart or sweet a flavour the recipe requires.

Nº 23

Stanhopea tigrina

upside down orchid
by Sarah Anne Drake from James Bateman
The Orchidaceae of Mexico and Guatemala,
1837–43

This orchid is endemic to Mexico and lives in the tree canopy as an epiphyte in tropical regions. Its large thick flowers hang downwards to attract visiting male orchid bees. These bees collect the scent of the flowers to make themselves more attractive to the opposite sex.

Nº 24

Dysphania ambrosioides

epazote

from François Pierre Chaumeton
Flore Médicale Décrite, 1815–20

A strongly fragrant herb used in many traditional dishes both for flavour and its medicinal qualities. It is thought to be anti-flatulent, but care must be taken not to eat too much.

No 25

Ipomoea tricolor

Mexican morning glory
by Marianne North from the
Marianne North Collection, Kew, 1882

This beautiful climber has bright blue flowers that will open at first light and slowly fade over the day, hence the name 'morning glory'. Each disk of petals features an origami star where the petals overlap.

Nº 26

Selenicereus undatus

dragon fruit, pitahaya, pitaya
from Ludwig Georg Karl Pfeiffer and Friedrich Otto *Abbildung und Beschreibung Blühender Cacteen,* 1843–50

This sizeable cactus grows by spreading its large, flattened stems over the ground and has been known to climb many metres up rocks or trees. The white flowers are pollinated by bats and produce a globular pink fruit with edible pulp.

Nº 27

Lantana camara

blacksage, lantana, yellow sage, cinco negritos
by S. Holden from J. Paxton *Magazine of Botany and Register of Flowering Plants*, 1843

Although this plant may be seen as a tenacious weed in many parts of the world, others regard it as a highly decorative garden plant that will flower all year round. The peculiar scent of its crushed leaves may allude to medicinal qualities. It is being investigated for a variety of potential benefits, including anti-inflammatory and anti-cancer activities.

Nº 28

Agave horrida

agave

by Harriet Thiselton-Dyer from
Curtis's Botanical Magazine, 1880

All parts of the agave are useful; from the sweet syrup within the leaves to the sharp spines that can be used as sewing needles. But it is the hearts of the blue agave, *Agave tequilana*, that provide the drink Tequila. Hundreds of millions of plants are individually selected to be harvested each year and will provide a different flavour depending on where they were grown. Agave from the valleys of the region of Tequila have an earthy terroir whilst those from higher altitudes tend to be sweeter.

Nº 29

Vachellia sphaerocephala

bee wattle, bull's horn thorn
by Harriet Thiselton-Dyer from
Curtis's Botanical Magazine, 1899

This shrub is endemic to Mexico and has a strong relationship with ants. Symbiotic relationships with ants, or myrmecophytism, is very common throughout the plant world. In return for shelter, ants will defend their territory from predators and may even provide food or disperse seeds.

Nº 30

Plumeria rubra

frangipani

from Edith Holland Norton *Brazilian Flowers Drawn from Nature in the Years 1880–1882 in the Neighbourhood of Rio de Janeiro*, 1893

These large ornate trees were traditionally very important to the native peoples of Mexico. The Aztecs regarded them as a status symbol, and large trees would adorn the gardens of nobles; while the Mayans regarded the flowers as a symbol of female sexuality and related the tree to the gods of fertility and life.

Nº 31

Penstemon roseus

penstemon, beardtongue
by Lilian Snelling, Kew Collection, 1951

Every flower within these colourful spires has a hairy inner surface to its protruding lower lip – hence the name 'beardtongue'. Although endemic to mountainous areas, these plants can be delicate during the winter months when planted away from home. Make sure the soil drains well and resist the urge to prune them before the cold has passed to ensure the best chance of survival.

Nº 32

Cucurbita pepo

squash

from Ernst Benary *Album Benary*, 1876–82

This squash species is thought to be the oldest domesticated plant in the world, with evidence from Oaxaca that could date back 10,000 years. This may go some way towards explaining how this one plant has so many completely different-looking cultivars. For example, this species has given us both the pumpkin, and the courgette.

Nº 33

Byrsonima crassifolia

nanche

by P. J. F. Turpin from *Aimé Bonpland Nova Genera et Species Plantarum,* 1815–25

The small yellow berries of *Byrsonima* are used in many candied desserts and ice creams. They can also be steeped in rum to create the drink licor de nanche.

N° 34

Euphorbia pulcherrima

poinsettia

from *Curtis's Botanical Magazine*, 1836

Poinsettia are familiar to many around the world but in their native tropical dry forest, wild plants can grow a lot larger than those adorning our homes at Christmas. Some are up to 4 m (13 ft) tall. The Aztecs cultivated poinsettia and used the plants as a red dye, as well as for various medicines, but these plants are now under threat due to deforestation.

Nº 35

Phaseolus vulgaris

bean

from L. Van Houtte *Flore des Serres et des Jardins de l'Europe*, 1849

This bean species is the source of around 18 cultivars, including the kidney bean, pinto and white bean. They would traditionally have been grown with maize and squash in a formation called the 'Three Sisters', with maize planted in a block, the bean climbing up the maize and squash covering the ground.

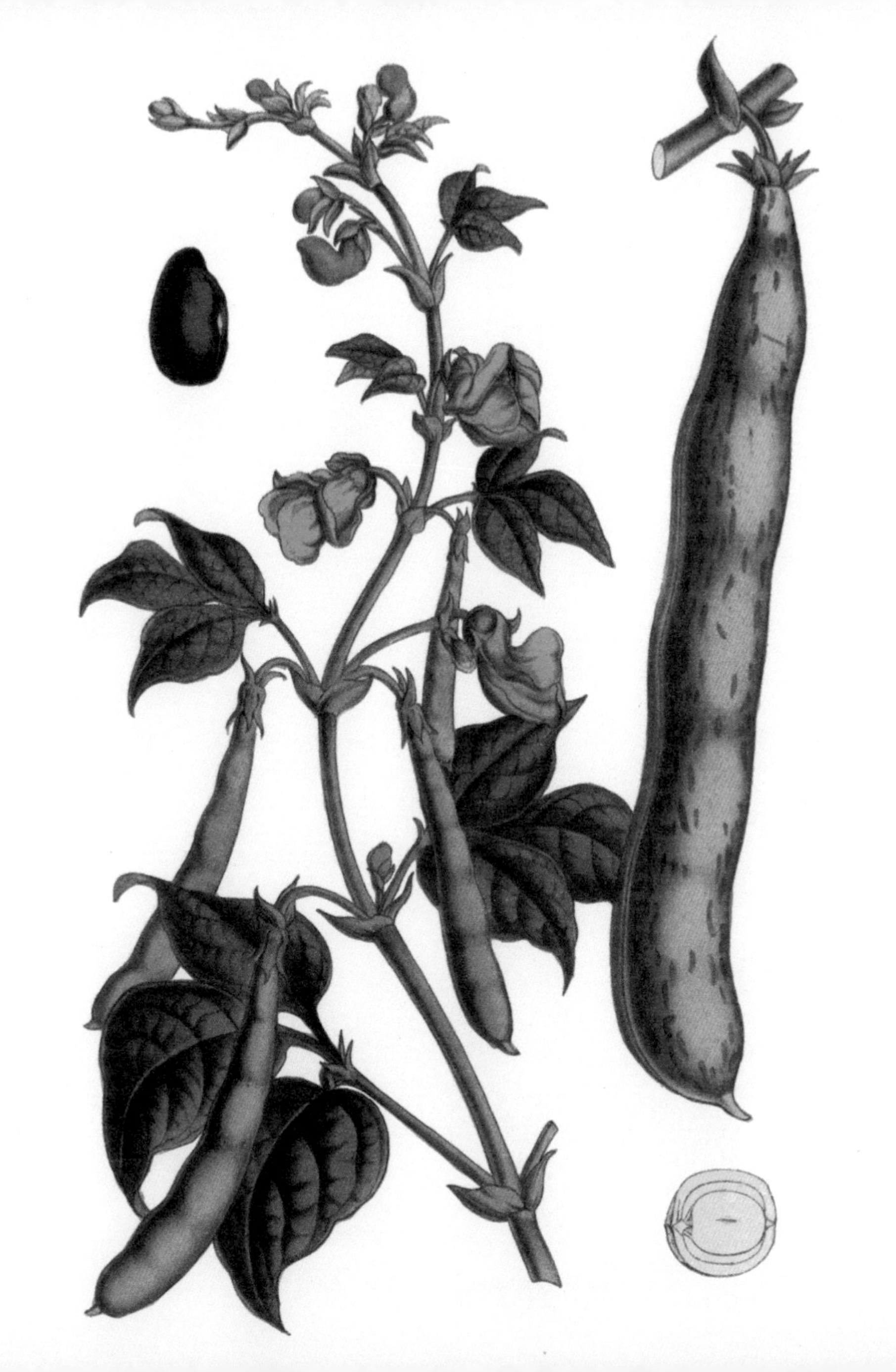

No 36

Zamia furfuracea

cardboard palm

from *Curtis's Botanical Magazine*, 1818

Zamia are a type of cycad, a group of plants that have been around since the dinosaurs. They have trunks and leaves like a palm tree, the leaves unfurl like a fern, they produce cones like a conifer and some even have surface roots that look like coral. Mexico has 70 species of cycad, of which 25 are endangered.

Nº 37

Dahlia species

dahlia

from C. V. D. d'Orbigny *DictionnaireUniversel d'Histoire Naturelle*, 1839–49

The national flower of Mexico, dahlias are found throughout the country and are world famous as an ornamental plant. The Aztecs used the tubers as an important food source, almost akin to a potato or yam, while the hollow stems of the tree dahlia (*Dahlia imperialis*) were used as pipes to carry water.

№ 38

Salvia leucantha

Mexican bush sage
by Walter Hood Fitch from
Curtis's Botanical Magazine, 1847

Native to the tropical conifer forests of Central and Eastern Mexico, this *Salvia* forms large spires of purple bracts which support and protect the flowers until they open. The large white flowers are then visited by butterflies, hummingbirds and bees.

No 39

Persea americana

avocado

by J. T. Descourtilz from *Flore Pittoresque et Médicale des Antilles*, 1821–9

The avocado hails from the highlands of central Mexico and forms the basis of many traditional dishes, including guacamole. It is now popular around the world and is regarded as a healthy food with many nutritional benefits.

Nº 40

Vanilla planifolia

vanilla

by Geneviéve de Nangis Regnault from François Regnault *La Botanique Mise à la Portée de Tout le Monde*, 1774

This succulent vine is where we get vanilla flavouring from. The creamy orchid flowers open from a small cluster of buds, only one a day, each lasting about 8 hours each. Once pollinated, each vanilla pod then takes 6–18 months to ripen. In Mexico, there are many pollinators that can do this job, including the orchid bee itself, but much of the vanilla we buy nowadays is grown around the world and must therefore be pollinated by hand.

ILLUSTRATION SOURCES

Books and Journals

Baillon, H. (1876–92). *Dictionnaire de Botanique.* Volume 4. Hachette, Paris.

Bateman, J. (1837–43). *The Orchidaceae of Mexico and Guatemala.* James Ridgway and Sons, London.

Benary, E. (1876–82). *Album Benary.* G. A. Koenig, Erfur.

Bonpland, A. (1815–25). *Nova Genera et Species Plantarum.* Volume 5. Sumtibus Librairie Graeco-Latino-Germanicae, Paris.

Carrière, E. A. (1889). *Capsicum annuum. Revue Horticole.* Volume 61.

Chaumeton, F. P. (1815–20). *Flore Médicale Décrite.* Volume 1. Panckoucke, Paris.

Descourtilz, M. É. (1821–9). *Flore Pittoresque et Médicale des Antilles.* Volume 8. Chez Corsnier, Paris.

Edwards, S. T. (1815–47). *Edwards' Botanical Register.* Volume 6. James Ridgway and Sons, London.

Figuier, L. (1867). *The Vegetable World.* Chapman & Hall, London.

Hayne, F. G. (1805–46). *Getreue Darstellung und Beschreibung der in der Arzneykunde Gebräuchlichen Gewächse.* F. G. Hayne, Berlin.

Hibberd, S. (1879–87). *Familiar Garden Flowers.* Cassell, London, Paris, Melbourne.

Hooker, J. D. (1871). *Tillandsia ionantha. Curtis's Botanical Magazine.* Volume 97, t. 5892.

Hooker, J. D. (1874). *Chamaedorea geonomiformis. Curtis's Botanical Magazine.* Volume 100, t. 6088.

Hooker, J. D. (1880). *Agave horrida. Curtis's Botanical Magazine.* Volume 106, t. 6511.

Hooker, J. D. (1889). *Tigridia pringlei. Curtis's Botanical Magazine.* Volume 115, t. 7089.

Hooker, J. D. (1899). *Acacia sphaerocephala. Curtis's Botanical Magazine.* Volume 125, t. 7663.

Hooker, W. J. (1834). *Helianthus speciosus. Curtis's Botanical Magazine.* Volume 61, t. 3295.

Hooker, W. J. (1836). *Poinsettia pulcherrima. Curtis's Botanical Magazine.* Volume 63, t. 3493.

Hooker, W. J. (1842). *Opuntia decumbens. Curtis's Botanical Magazine.* Volume 68, t. 3914.

Hooker, W. J. (1847). *Salvia leucantha. Curtis's Botanical Magazine.* Volume 73, t. 4318.

Jacquin, J. F. von. (1811–44). *Eclogae Plantarum Rariorum aut Minus Cognitarum*. J. F. von Jacquin, Vindobona.

Jacquin, N. J. von (1780–1). *Selectarum Stirpium Americanarum Historia*. N. J. von Jacquin, Vindobona.

Köhler, F. E. (1887). *Köhler's Medizinal-Pflanzen*. Volume 2. F. E. Köhler, Gera-Untermhaus.

Norton, E. H. (1893). *Brazilian Flowers Drawn from Nature in the Years 1880–1882 in the Neighbourhood of Rio de Janeiro*. Coombe Croft, Kingston upon Thames, Surrey.

d'Orbigny, C. V. D. (1839–49). *Dictionnaire Universel d'Histoire Naturelle*. Volume 3. A. Pilon, Paris.

Paxton, J. (1843). *Magazine of Botany and Register of Flowering Plants*. Volume 10.

Pfeiffer, L. G. K. and Otto, F. (1843–50). *Abbildung und Beschreibung Blühender Cacteen*. Volume 1. T. Fischer, Cassel.

Phillips, J. (1842). *Mexican Scenery*. Ackermann and Co., London.

Regnault, F. (1774). *La Botanique Mise à la Portée de Tout le Monde*. F. Regnault, Paris.

Sims, J. (1813). *Cosmea bipinnata. Curtis's Botanical Magazine.* Volume 37, t. 1535.

Sims, J. (1818). *Zamia furfuracea. Curtis's Botanical Magazine.* Volume 45, t. 1969.

Sweet, R. (1823–38). *The British Flower Garden*. Volume 1. James Ridgway and Sons, London.

Thomé, O. W. (1886–89). *Flora von Deutschland Österreich und der Schweiz*. Volume 1. F. E. Köhler, Gera-Untermhaus.

Van Houtte, L. (1849). *Phaseolus vulgaris. Flore des Serres et des Jardins de l'Europe*. Volume 5, p. 433.

Van Houtte, L. (1873). *Echeveria agavoides. Flore des Serres et des Jardins de l'Europe*. Volume 19.

Wight, R. (1840–50). *Illustrations of Indian Botany*. J. B. Pharoah, Chennai.

Wilhelm, G. T. (1810–11). *Unterhaltungen aus der Naturgeschichte*. Engelbrechtschen Kunsthandlung, Augsburg.

Art Collections

Marianne North (1830–90). Comprising over 800 oils on paper, showing plants in their natural settings, painted by North, who recorded the world's flora during travels from 1871 to 1885, with visits to 16 countries in 5 continents. The main collection is on display in the Marianne North Gallery at Kew Gardens, bequeathed by North and built according to her instructions, first opened in 1882.

FURTHER READING

Mills, Christopher. (2016). *The Botanical Treasury*. Welbeck Publishing, London in association with the Royal Botanic Gardens, Kew.

North, Marianne and Mills, Christopher. (2018). *Marianne North: The Kew Collection*. Royal Botanic Gardens, Kew.

Payne, Michelle. (2016). *Marianne North: A Very Intrepid Painter*, revised edition. Royal Botanic Gardens, Kew.

Willis, Kathy and Fry, Carolyn. (2014). *Plants from Roots to Riches*. John Murray, London in association with the Royal Botanic Gardens, Kew.

Online

www.biodiversitylibrary.org – biodiversity and natural history literature including many rare books.

www.kew.org – information on Kew's science, collections and visitor programme.

www.plantsoftheworldonline.org – authoritative information on the world's flora from the botanical literature.

ACKNOWLEDGEMENTS

Kew Publishing would like to thank the following for their help with this publication: in Kew's Library and Archives Fiona Ainsworth, Anne Marshall and Cecily Nowell-Smith; for digitisation work, Paul Little.

INDEX

First published in 2022
Royal Botanic Gardens, Kew,
Richmond, Surrey, TW9 3AB, UK
www.kew.org

ISBN 978 1 84246 767 1

Distributed on behalf of the Royal Botanic Gardens, Kew in North America by the University of Chicago Press, 1427 East 60th St, Chicago, IL 60637, USA.

British Library Cataloguing in Publication Data
A catalogue record for this book is available from the British Library

Design: Ocky Murray
Page layout: Christine Beard
Production Manager: Jo Pillai
Copy-editing: Ruth Linklater

Printed and bound in Italy by Printer Trento srl.

Front cover image: *Dahlia* species (see page 84).

Endpapers: Las Peñas Cargadas from John Phillips *Mexican Scenery*, 1842.

p2: Mexican Plants from Louis Figuier *The Vegetable World*, 1867.

p4: *Zea mays*, corn, maize, from Gottlieb Tobias Wilhelm *Unterhaltungen aus der Naturgeschichte*, 1810–11.

p9, 94: *Capsicum annuum*, chilli pepper, pimiento, chiplín, by P. de Longpré from E. A. Carrière *Revue Horticole*, 1889.

p10–11: The Cave of Cuernavaca from John Phillips *Mexican Scenery*, 1842.

For information or to purchase all Kew titles please visit shop.kew.org/kewbooksonline or email publishing@kew.org

Kew's mission is to understand and protect plants and fungi, for the wellbeing of people and the future of all life on Earth.

Kew receives approximately one third of its funding from Government through the Department for Environment, Food and Rural Affairs (Defra). All other funding needed to support Kew's vital work comes from members, foundations, donors and commercial activities, including book sales.

Publishers note about names

The scientific names of the plants featured in this book are current, Kew accepted names at the time of going to press. They may differ from those used in original-source publications. The common names given are those most often used in the English language, or sometimes vernacular names used for the plants in their native countries.

OTHER TITLES IN THE
KEW POCKETBOOKS SERIES

Cacti

Palms

Honzō Zufu

Japanese Plants

Carnivorous Plants

Wildflowers

Fungi

Festive Flora

Herbs and Spices

Fruit

Orchids